VENUS FLYTRAP

ASHLEY GISH

VENUS FLYTRAP

CREATIVE EDUCATION • CREATIVE PAPERBACKS

Published by Creative Education
and Creative Paperbacks
P.O. Box 227, Mankato, Minnesota 56002
Creative Education and Creative Paperbacks are imprints of
The Creative Company
www.thecreativecompany.us

Design and art direction by Blue Design
Edited by Kremena Spengler

Photographs by Alamy Stock Photo/Maximum Film, 23; Dreamstime/Bogdan Lazar, 1, Sikth, 16, Verastuchelova, cover, 15; Getty Images/Ed Reschke, 2, Joao Paulo Burini, 13, Paul Starosta, 17, 21, Oxford Scientific, 9; Microsoft AI/23; Unsplash/Bill Nino, 5; Wikimedia Commons/Björn S., 21, Calyponte, 6–7, Judy Gallagher, 14–15, Mucrow, 18–19, Noah Elhardt, 14, Rosťa Kracík, 11, 21, Sanjay Acharya, 10, Tippitiwichet, 20, 21, User:Citron, 22
Every effort has been made to contact copyright holders for material reproduced in this book. Any omissions will be rectified in subsequent printings if notice is given to the publisher.

Library of Congress Cataloging-in-Publication Data
Names: Gish, Ashley, author.
Title: Venus flytrap / Ashley Gish.
Description: Mankato, Minnesota : Creative Education and Creative Paperbacks, [2026] | Series: Look! | Includes bibliographical references and index. | Audience: Ages 6–9 | Audience: Grades 2–3 | Summary: "A carnivorous plant with jaw-like leaves and toothlike spikes that trap insects? Elementary-level readers will learn all about the Venus flytrap with this striking introduction to the unusual plant's life cycle"– Provided by publisher.
Identifiers: LCCN 2024044507 (print) | LCCN 2024044508 (ebook) | ISBN 9798889895893 (library binding) | ISBN 9781682777558 (paperback) | ISBN 9798889896692 (ebook)
Subjects: LCSH: Venus's flytrap–Juvenile literature.
Classification: LCC QK495.D76 G57 2026 (print) | LCC QK495.D76 (ebook) | DDC 583/.75–dc23/eng/20241223
LC record available at https://lccn.loc.gov/2024044507
LC ebook record available at https://lccn.loc.gov/2024044508

TABLE OF CONTENTS

There are 750 kinds of **carnivorous** plants on Earth. One is the Venus flytrap. Its leaves look like jaws. They have spikes. The spikes look like teeth. They trap insects inside.

The Venus flytrap grows in wetlands in North and South Carolina. It is against the law to dig one up. It's becoming rare in the wild.

carnivorous: a living thing that eats meat

The Venus flytrap was named Venus after the Roman goddess of love.

10

Venus flytrap flowers grow on long stalks. Insects look for food in the flowers. The insects carry **pollen** to the inside of the flower. This is where seeds grow.

In a few weeks, the flowers dry up. The seeds fall to the ground. They begin to sprout. The seeds become **seedlings**. After a few years, they make flowers of their own.

pollen: yellow powder made by plants that is used to make other plants

seedling: a young plant

Venus flytraps can also grow from leaves and **rhizomes** (RY-zomes). If a leaf falls off the plant and touches the ground, roots will begin to grow from it. It will grow into a whole new plant! The same is true of rhizomes.

The Venus flytrap will grow and spread a lot during its lifetime. In fact, many new plants will grow from the rhizomes of just one Venus flytrap.

rhizome: an underground stem from which a new plant can grow

Each trap can close up to 12 times before it dies and falls off.

The plant's jawlike traps have sweet liquid inside. It's called nectar. Insects come to drink it. As soon as the insects' feet touch the fuzzy insides of the trap, the trap snaps closed.

Only movement can make the trap shut. Usually, it's a wiggling insect. The trap seals shut tightly. Nothing can get out. Nothing can get in. The trap is like a mouth and stomach in one.

The inside of the trap makes juices. They break down the insect's soft body. But they can't break its hard outer shell. When the trap opens, all that's left is the insect's **exoskeleton**.

The Venus flytrap gets most of its energy from the insects it eats. Sometimes it eats flies. It also eats spiders, beetles, and ants.

exoskeleton: a hard covering that protects the outside of a body

It takes seven to ten days for the Venus flytrap to eat a trapped bug.

The Venus flytrap has no plant family. But some plant lovers have grown Venus flytraps to look different from each other.

Wacky Traps are red and yellow. They have strange teeth. Sawtooth Venus flytraps have short, spiky teeth. Big Mouths are some of the biggest Venus flytraps. Their bright red traps grow close to the ground.

Big Mouth
Venus flytrap 19

20

Venus flytraps may live up to 20 years.

A Venus flytrap grows from a seed, leaf, or rhizome. Roots grow down into the soil. Stems and leaves grow up into the air. The trap opens. Insects crawl or fly in. The Venus flytrap gets energy from the insects. In a few years, flowers will grow. More Venus flytraps come from the parent plant.

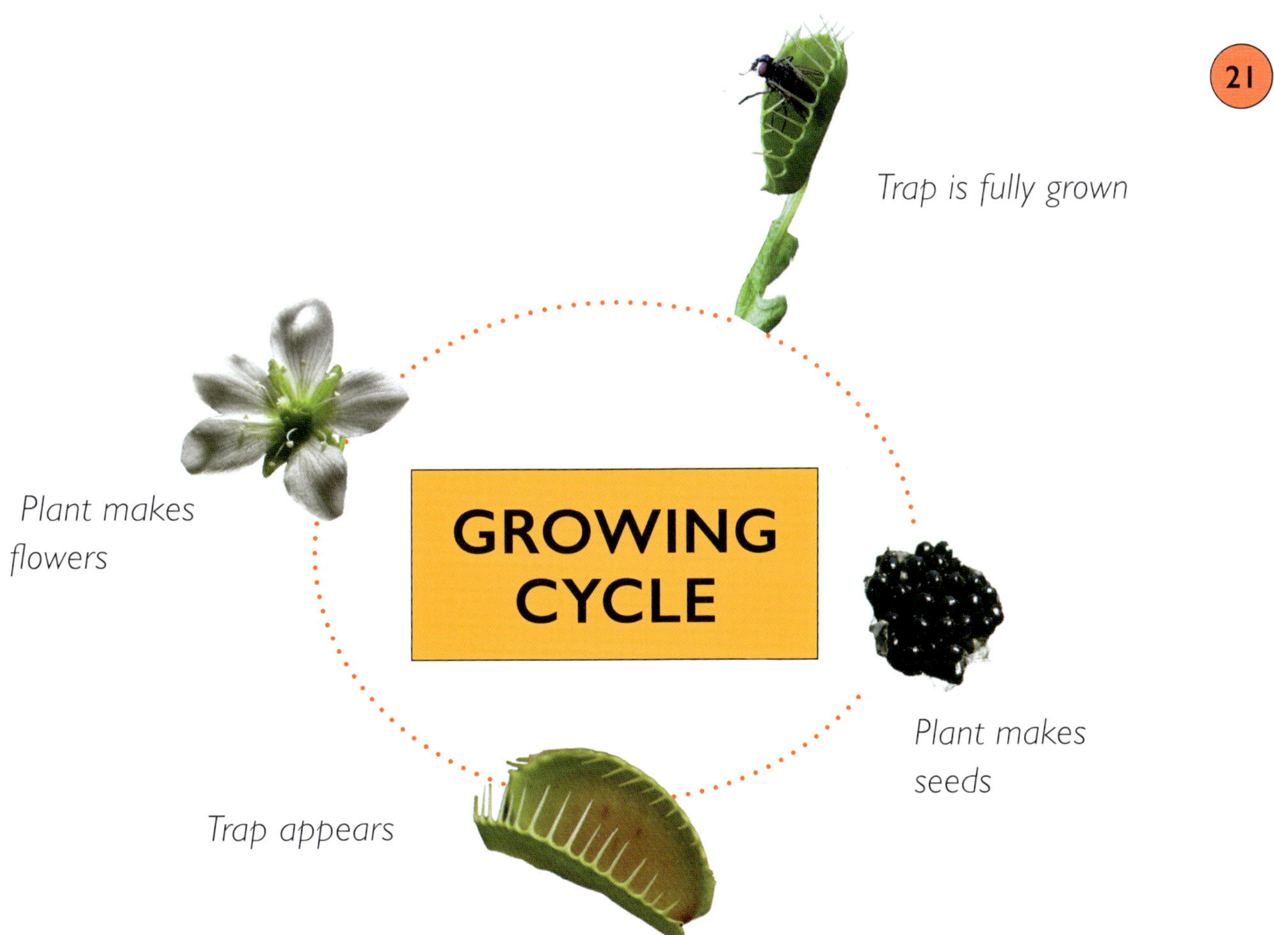

Many movie monsters are based on Venus flytraps. The best known is Audrey II from *Little Shop of Horrors* (1986). It is a cross between a butterwort and a Venus flytrap. It is always hungry! And in 1989, Godzilla fought a monster called Biollante, which looked like an enormous Venus flytrap.

WEBSITES

Awesome 8 Carnivorous Plants
https://kids.nationalgeographic.com/awesome-8/article/carnivorous-plants
Learn more about carnivorous plants, including the Venus flytrap.

Venus Flytrap
https://scaquarium.org/our-animals/venus-flytrap
Learn about Venus flytraps on the South Carolina Aquarium's website.

READ MORE

Braun, Eric. *Corpse Flower vs. Venus Flytrap*. Mankato, MN: Black Rabbit Books, 2018.

Griffin, Mary. *Fantastic Plants: Meat-Eating Venus Flytraps*. New York: PowerKids Press, 2023.

Markovics, Joyce. *Carnivorous Plants*. Ann Arbor, MI: Cherry Lake Publishing, 2021.

INDEX